Ruby's Room

Photographic Miniatures
by Bettina von Zwehl

Exhibition list

1 — *Ruby (with Diamond),* 2012
C-type, image 6.4 × 7.6 cm

2 — Necklace
In collaboration with Laura Lee Jewellery
Eye portrait (Ruby), 2013
C-type print surrounded by old cut diamonds, image 2.4 × 1.8 cm

3 — *Ruby,* 2013
C-type, image diameter 5.7 cm

4 — *Self-portrait (for Ruby's Room),* 2013
C-type, image diameter 5.7 cm

5 — Ring
In collaboration with Laura Lee Jewellery
Eye portrait (David), 2013
C-type print set in 18ct gold ring, image 2.2 × 1.7 cm

6 — *Golden Eye (Ruby),* 2012
C-type, image oval 2.9 × 4.4 cm

7 — *Eye portrait (Katherine),* 2012
C-type, image oval 2.9 × 4.4 cm

8 — *Eye portrait (Lauren),* 2012
C-type, image oval 2.9 × 4.4 cm

9 — *Eye portrait (Madeleine),* 2012
C-type, image oval 2.9 × 4.4 cm

10 — *Eye portrait (Molly),* 2012
C-type, image oval 2.9 × 4.4 cm

11 — *Eye portrait (Sagal),* 2012
C-type, image oval 2.9 × 4.4 cm

12 — *Eye portrait (Clemmie),* 2012*
C-type, image oval 2.9 × 4.4 cm

13 — *Eye portrait (Bradley),* 2012*
C-type, image oval 2.9 × 4.4 cm

*Not illustrated

Introduction

Alexander Sturgis, Director
Holburne Museum

When we invite contemporary artists into the Holburne to work in our building or with our collection, we hope that both we and the artist will be surprised by the experience. An ideal is that artists encounter something at the museum that creates a spark that lights a fire that burns in new and unexpected ways.

This book and the exhibition it accompanies are evidence of just such an encounter. Bettina von Zwehl has long been concerned with the conundrums, conventions and paradoxes of the portrait and so was in many ways a natural partner for the Holburne with its particular strength in portraiture. Her work exploring the miniature photographic portrait also suggested a potential affinity with a collection that delights in the small and the precious. But although we might have been confident that we would 'get on', none of us predicted that a single miniature in our displays would lead to such a rich new seam of work and thought. In retrospect maybe we should have done. The eye portrait, in the tensions it sets up between the rational and the emotional, the intimate and the scientific, the viewer and the viewed chimes absolutely with Bettina von Zwehl's long-held preoccupations. I hope that the fire lit by the encounter at the Holburne will continue to glow hot for some time to come.

There are many people to thank: Bettina herself of course but also Katy Barron for making the introduction and recognising that something good might come of it; Dr Hanneke Grootenboer for deepening our collective understanding of the eye portrait; and Laura Lee for her important role in turning Bettina's images into objects. Thanks too are due to the Victoria Art Gallery, Bath for the generous loan of the eye-miniature that provided the vital spark. Finally, huge and repeated thanks to Arts Council England for their crucial support of both the exhibition and this publication and who consistently help sparks become fires.

Private View

Dr Hanneke Grootenboer
University of Oxford

Bettina von Zwehl's (b.1971, Munich, Germany) bejeweled necklace pendant holds an unexpected surprise. Bending over it, we find ourselves confronted with the brown iris of an eye, calmly staring at us as if from a peephole (see **2**). Set in a gold frame encrusted with old-cut diamonds, the eye is in fact a hybrid – part jewel, part miniature, not a photograph in the usual sense, and not quite a portrait. Radically cropped so that we see nothing but a single watchful eye, the image presented makes it impossible to identify its sitter. Indeed, this eye's identity will remain forever hidden to everyone but the jewel's owner. Aside from the valuable diamonds, this token's significance lies in its owner's exclusive knowledge of to whom the eye belongs. The necklace is long, and, when worn, the pendant hangs low on the chest, close to the heart. What is most extraordinary about this highly personalized piece is that it treasures not the likeness (as in other portrait miniatures) but the *gaze* of a loved one as it is rests upon the lover's body. What does the eye wish to see? What is its concern?

Exceptional as this art work may be, it takes its place within a rich tradition that reaches back to the end of the eighteenth century. There began at that time a craze for miniature portraits of single eyes that were mounted in rings, pendants, or brooches, and were exchanged among the upper class between lovers and family members. The Holburne Museum exhibits a rather unusual example of so-called eye miniatures (see fig.1): from an oval display not one but five miniature eyes stare out at the viewer! This unique souvenir – there is literally no other like it – is made of ivory and surrounded by purple-coloured gemstones; it is mounted in a steel frame as a pendant. It is as if we are looking at a bundle of stares, a bouquet of gazes resting upon us.

If we look closer we see that each eye is in fact a highly individualized mini-portrait of one of five distinct sitters'

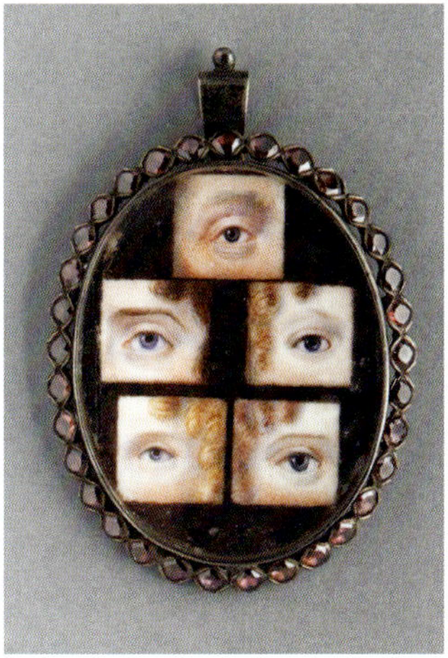

fig.1
Family of Eyes
British, about 1810
Watercolour on ivory
Height 3.8 cm
Victoria Art Gallery, Bath

faces. While there is very little to go on within these five frames, the facial parts are so delicately rendered that each is imbued with a sense of individuality and character. The eyes give clear indications of each sitter's gender and age. The eye on top shows a slight hint of a grey-speckled sideburn. In contrast to the other eyes, all of are framed by flowing locks of hair or fine curls, this one must belong to an elderly man. The rather large, bright blue eye, with a finely shaped eyebrow on the second row to the left, is a young woman's, while the other three eyes are clearly more childlike, the one on the lower left being much smaller in size, and probably belonging to a very young child. We seem to be looking here at an ocular family portrait, perhaps of a parent with four of his children created to commemorate a special occasion. To be able to hold gazes of five loved ones in one's hand, and to feel their eyes resting upon you wherever you would go, must have given its owner a special feeling of love and reassurance. Surely, these five eyes are there not just to be seen, but *to see*; they watch their owner's every move, as if a kind of camera *avant la lettre*, a channel through which the family members could stay in contact with the gift's recipient, over distance. It is as if this memento communicates the message 'we will look out for you,' expressing a continuous concern for the beholder's wellbeing. These are 'speaking gazes' reaching out to the addressee, almost as a mode of touch.

This is exactly the kind of intimacy invoked by von Zwehl's jewel. Whereas eye portraits were set in rings or brooches, they were usually not worn openly. Although the fashion was for women to wear portrait miniatures of their husbands pinned to their bosoms or mounted as bracelet clasps, it was considered unmanly for gentlemen to walk around with a female portrait on the body. Men, therefore, wore miniatures on ribbons under their clothes, or treasured portraits of their beloveds in the privacy of their study, often keeping them in writing boxes.

Eye portraits were considered to be for private viewing only. The ocular group portrait we see here was probably never worn openly, and could well have been pinned to the inside of the lining or facing of a dress, or inside compartment, to be uncovered only in solitude. The owner of the portrait would experience less the pleasure of seeing a picture of a loved one and more the sheer joy of *being seen*, and to imagine the special kind of touch of being perpetually watched by those with whom you are in love.

This sort of circuit of intimate vision that was effected by eye portraits, where you and your loved one could be alone together despite physical separation, fell into oblivion for centuries – and now has been restored by von Zwehl. Her eye portraits in fact continue a tradition for objectified affection whereby a gaze is made as palpable as it is portable. Intimacy can be defined in regard to the gaze as a space where one would be sheltered from public exposure, exposed to no voyeuristic eye. Von Zwehl's images evoke this kind of extreme privacy. Comparing the children's eyes in the ocular family portrait with the hazelnut eye in the diamond-studded pendant, we suddenly become aware of the brown eye's purity and innocence. This is a child's eye resting upon us. To be touched by this relatively untouched gaze reminds us of what it means to look out for someone, openly, without inhibitions or constraints.

Approaching this bejeweled photograph, and others like it, by bending over to look at it evokes the sense of entering a different realm. Demanding very little, this small eye portrait subverts our usual sense of the subject and object of looking. If, as we have seen in eighteenth-century literature, a powerful gaze can 'touch' over distance, and speak countless words, here we find that it can also bring silence. In drawing nearer to the tiny picture, we are silenced. It cancels out the rest of the world, and, for a split second, there is a genuine encounter, a clash of our gaze and that of the portrayed one. The eye does not seem to *belong* to anyone but us, and its serene stare transforms us, as viewers, into its exclusive focus: the cause for its concern.

It has been argued that the invention of photography killed portrait miniatures. Many miniaturists were forced to replace their fine brushes and pieces of ivory with a camera, backdrop, and head stand so as to keep their clientele. However, as von Zwehl's eye portraits so aptly demonstrate, photography did not kill miniature art. Portraiture's essence – its watchfulness and its concern – lives on, and continues to draw us in so that we might be seen.

1 — *Ruby (with Diamond)*, 2012

2 — Necklace
Eye portrait (Ruby), 2013

Preceding spread, left to right:

3 — *Ruby*, 2013

4 — *Self-portrait (for Ruby's Room)*, 2013

Opposite:

5 — Ring
Eye portrait (David), 2013

Ruby's Room

Katy Barron
Exhibition Curator

When I invited Bettina von Zwehl on a visit to the Holburne Museum in late 2011 I hoped that she might want to make new work in response to the museum's miniature collection. The Holburne has a small group of mostly eighteenth and nineteenth century miniatures, which includes a tiny *Family of Eyes* (see fig.1) on loan to the Holburne from the Victoria Art Gallery, Bath. It was this unusual work of art that attracted Bettina, who began to research the history of the eye miniature. Bettina has been making photographic miniatures since her residency at the Victoria and Albert Museum in 2011. Whilst there, she produced *Made Up Love Song* (fig.2) – a series of 34 photographic miniature profile portraits of a gallery assistant, made at different times of day but always in the same place, over 6 months. The eye miniatures that she began to make after her visit to the Holburne are an extension of this work and resulted in the exhibition *Ruby's Room*. The exhibition falls into two parts – the single framed eyes of school children, which have been placed within the permanent galleries throughout the Holburne Museum and the works in the Wirth Gallery, which focus on von Zwehl's daughter Ruby.

Von Zwehl has always made photographs of people. Her earlier series focused on an exploration of the way that her subjects appeared whilst undergoing different physical and emotional experiences such as being woken from a deep sleep or listening to a piece of music in the dark. *Made Up Love Song* was a departure for von Zwehl in that the miniatures are a long-term examination of the changes, both physical and emotional, of an individual, someone with whom the artist developed a connection. The works in the Wirth Gallery (see 1 to 5) present us with even closer relationships as they record the artist as well her daughter and her husband, each of whom are identified for the viewer – another departure for von Zwehl.

fig.2
Made Up Love Song, part 7, 2011
C-type, image diameter 5.8 cm

Bettina's initial project for photographing eyes in miniature was to record the eye of every child at her daughter's primary school. The sheer volume of children to be photographed meant that she had little time to get to know each child and the resulting works have an indexical quality when we first see them. However each eye portrait offers more material the more closely we examine it – we begin to guess at the age, gender, race and even temperament of the sitter. Tiny differences between the photographs – such as the edge of a headscarf or the curve of a brow become hugely important, as each child's eye gradually emerges as that of an individual. It is notable that many of the children in the Primary School were able to find their own eyes from amongst the whole school group. When encountered unexpectedly in a gallery, a place designed for looking, the children's eyes seem to confront us, almost challenging us to look away and it is unclear who is looking at whom.

The framed wall-mounted works embody little of the intense personal meaning of painted eye miniatures, which traditionally were carried or worn, but rather reflect the huge power of the photographic image, even when drastically reduced in size. This is particularly apparent when the visitor encounters an image of Ruby's eye, framed in gold, which has been placed within the main picture gallery of the museum. The contrast between Ruby's tiny brown eye and the large painted portraits is remarkable. The piece demands that the viewer consider the validity of the abstracted eye as a portrait and photography's relationship with painting. The invention of photography has often been cited as the death knell of the miniature portrait.

Photography and the eye are inevitably entwined, although the eye alone has rarely been a subject for fine art photographs. Two works made more than half a century apart both use the watchful eye as a proxy for a person.

Man Ray's *Object to be Destroyed* (see fig.3) is a ready-made piece incorporating his lover Lee Miller's eye – 'From atop a metronome, a single photographic eye stares out at us. It is a lovely eye .... Set into motion, however, the eye aggresses, disturbs and taunts. Ticking back and forth, it regulates and controls us .... With nerves frayed and frustration pent up .... We pick up the hammer, smash the metronome to bits; .... There is a moment of release, even satisfaction, ... But then we notice that ... the eye still stares, placidly, oblivious

fig.3
Metronome (Object to be Destroyed),
about 1923–72
Man Ray (1890-1976)
Wood, metal and paper
23 × 11 × 11 cm
Hamburger Kunsthalle, Hamburg

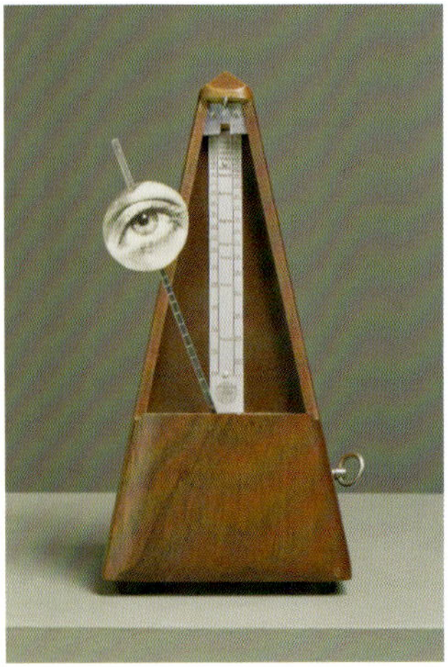

to the destruction wrought in its wake.'

In 1996 Alfredo Jaar (b.1956-) used a photograph of a pair of eyes in *The Eyes of Gutete Emerita,* a work about the Rwandan genocide. In it, the eyes of a young Tutsi woman who survived the massacre of her family are briefly illuminated within two lightboxes. The image is replaced with a series of texts about the massacre, but it is the brief appearance of Gutete Emerita's eyes that force the viewer to visualize the massacre through her eyes. Jaar was concerned that the endless flow of press images from Rwanda had led to a visual desensitization amongst the public and this piece was an attempt to force the viewer 'eye-to-eye' with the awful events described.

Von Zwehl's eye miniatures, like the works by Man Ray and Alfredo Jaar, represent the subject and also allow the viewer to see the world 'through their eyes'. The anonymity of the school children's eyes is contrasted with the intensely personal eyes and portraits presented in the cabinet-like Wirth Gallery. Viewing the five small works feels private (the scale of the work itself demands it) and invites parallels with the intimate encounter of a painted eye miniature. Bettina's daughter Ruby is the thread running through the works. She is first encountered in a profile portrait in which her eyes are closed, in contrast to every other eye in the exhibition. The small box in her hands is also shut, leaving the viewer to guess at its contents, and presenting Ruby as removed, perhaps dreaming. This mix of intimacy and distance imbues all of the works in the room. As with the *Family of Eyes* miniature, we are given such close access, and yet the interior lives of the sitters remains mysterious.

6 — *Golden Eye (Ruby)*, 2012

7 — *Eye portrait (Katherine),* 2012

8 — *Eye portrait (Lauren)*, 2012

9 — *Eye portrait (Madeleine)*, 2012

10 — *Eye portrait (Molly)*, 2012

11 — *Eye portrait (Sagal)*, 2012

Published to accompany the exhibition

Ruby's Room
Photographic Miniatures by Bettina von Zwehl

The Holburne Museum
22 May to 1 September 2013

The Holburne Museum
Great Pulteney Street
Bath BA2 4DB
T: 01225 388 588
www.holburne.org
Registered charity no. 310288

ISBN 9780903679039

Design: Northbank, Bath

LOTTERY FUNDED
Supported using public funding by
ARTS COUNCIL ENGLAND

Supported by: Laura Lee Jewellery

THE HOLBURNE MUSEUM